The Development of U.S. Missiles During the Space Race With the U.S.S.R.

Including a First Person Narrative
by
ROBERT TATA

Former Flight Test Engineer
Cape Canaveral, Florida

authorHOUSE®

AuthorHouse™
1663 Liberty Drive
Bloomington, IN 47403
www.authorhouse.com
Phone: 1-800-839-8640

First published by AuthorHouse 10/3/2011

ISBN: 978-1-4567-4084-9 (e)
ISBN: 978-1-4567-4083-2 (sc)

Library of Congress Control Number: 2011904251

Printed in the United States of America

Overview

THIS PUBLICATION PROVIDES A SHORT history of introductory information on ballistic missiles developed by the United States military including reference to the "Space Race" competition between the United States and Russia during the "Cold War". It features easily understood insight into the operation of rocket engines and the early stages of development of the Atlas Intercontinental Ballistic Missile System (ICBM), the first that was developed by the U.S. It also describes other U.S. missile programs and their deployment across the United States and Europe to combat the Russian threat. It concludes with a sometimes serious, sometimes humorous narrative by the author of his life as a Flight Test Engineer at Cape Canaveral, Florida. He describes his experiences working on the Atlas ICBM test stands and taking part in Atlas missile launches during its early stages of development. This publication was written for the interest and enjoyment of the average reader who might be interested in the "space race" through the eyes of a young engineer, fresh out of college, who had a small role in it. It includes eleven illustrations, several of which are full size photographs of Cape Canaveral missiles and missile launches from the author's personal collection.

Contents

Background

During World War II the United States and Russia fought as allies against Germany and Japan. After the war, the U.S., tired of the ravages of international conflict, greatly reduced its armed forces while Russia kept the bulk of its huge army active and forced Eastern European nations to join the Communist bloc. The term "Cold War" was used to describe the relationship between the U.S. and Russia because, although they regarded each other as enemies, there was no open warfare between the two. Behind the scenes both governments engaged in intense scientific research in the development of nuclear weapons and the means of delivering them over the vast distances that lie between the two countries. It was for this reason that in October 1945 the U.S. Army Air Force requested industry proposals for missile systems to deliver warheads as far as 6000 nautical miles (5200 statute or land miles).

Following the presentation of information on ballistic missiles, there is included a first person account by the author of his sometimes serious, sometimes humorous experiences as a Flight Test Engineer at Cape Canaveral, Florida on the Atlas Intercontinental Ballistic Missile System during its early stages of development.

Introduction

THIS PUBLICATION IS INTENDED TO provide introductory information on ballistic missiles developed by the United States military. The flight of a ballistic missile resembles that of an artillery piece where the projectile is given an initial thrust out the barrel and then coasts in an arced trajectory to its target. A ballistic missile used for military purposes is given an initial thrust from its engines and then coasts to its target in a similar curved path. Initially, U.S. missiles were designed after the German V2 rockets whereby a kerosene-like fuel and a liquid oxygen oxidizer were used as the propellants. Later, more advanced liquid and solid propellants were developed. The long range ICBMs have the capability of reaching most targets anywhere on the globe from bases in the U.S. The Medium Range Ballistic Missiles (MRBM) have the capability of hitting enemy targets from bases located in friendly countries in Europe. Short Range Ballistic Missiles (SRBM) were issued to combat troops for use on the battlefield. There is good reason to believe that ballistic missiles had a large part in preventing the outbreak of World War III.

Rocket Engines

A ROCKET IS AN ENGINE that produces more power than any other engine known. The word "rocket" is also used to describe the vehicle that is powered by a rocket engine. The Chinese used rockets against enemy soldiers in the 1200s. In the war of 1812, British soldiers used rockets to attack Fort McHenry, Maryland. Francis Scott Key while watching the battle wrote "the rockets' red glare" which is contained in "The Star-Spangled Banner", the United States National Anthem. "Missile" is also used to describe the vehicle that is powered by a rocket engine. Missiles are used for military purposes as well as to carry people and scientific equipment into outer space for peaceful means.

Rocket engines are internal combustion engines that rely on the production of high pressure gases from the chemical reaction of a fuel component and an oxidizer component to provide propulsive force. There are two kinds of rockets that are used on most missiles and spacecraft today: liquid propellant and solid propellant. Liquid propellants require proportionately large tanks, pumps, and plumbing systems to store and deliver the contents to a combustion chamber where the fuel and oxidizer are mixed and ignited by an electric spark. In the case of hypergolic (self-igniting) liquid propellants, the two components ignite on contact. Solid propellants have the fuel and oxidizer mixed in granular form. The grains are packed inside a cylindrical casing that has a hollow core extending down the center where the combustion takes place. The high temperature gases from the combustion of the grains flow into a nozzle which converts high temperature, high pressure gas into very high velocity gas-the reactive force of which propels the vehicle. (See Figures 1 & 2.)

Figure 1
Schematic
Rocket Engine
Liquid Propellant

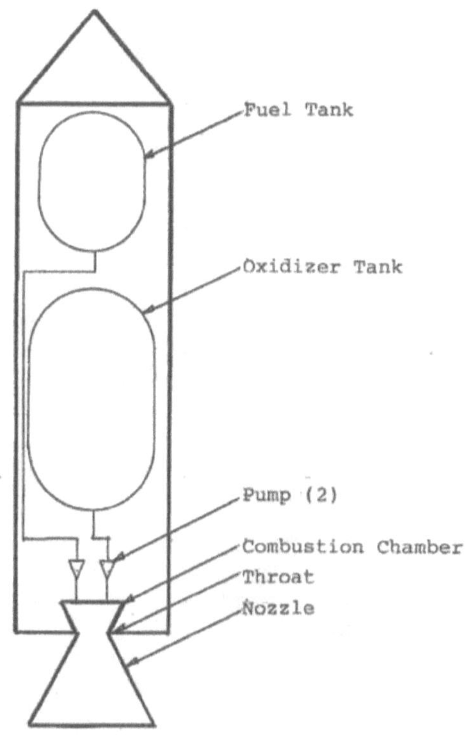

Figure 2
Schematic
Rocket Engine
Solid Propellent

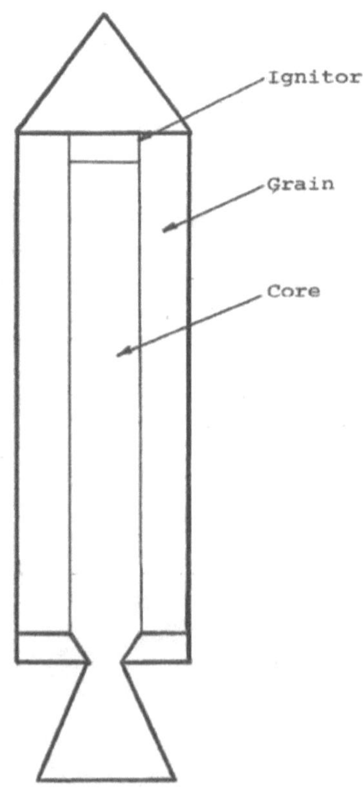

Rocket Engine Propellants

Most U.S. missiles use RP1 and liquid oxygen (LOX) as the propellants, the same used in the German V2 rockets. RP1 is similar to kerosene which is distilled from crude oil and is a mixture of various hydrocarbons. Hydrocarbons, as the name suggests, are hydrogen-carbon compounds which when mixed with LOX and ignited produce heat, carbon dioxide, and water. Liquid oxygen is obtained from distilled liquid air. Air is liquefied by lowering its temperature to minus 310F. The most powerful liquid propellant combination is liquid hydrogen (LH2) and LOX. Bulk quantities of LH2 are produced from natural gas. LH2 and LOX burn clean giving off only water and heat as the by-products. The chemical equation is:

$$2H_2+O_2-->2H_2O+heat$$

This combination results in an extremely large amount of high temperature gases making it the most powerful of all propellants used today. The missile that put the men on the moon and the space shuttle both use this combination. It was first pioneered on the Centaur upper stage vehicle which was previously used on many spacecrafts.

An advantage of liquid propellants is that engines can be stopped and restarted in space while some of the more advanced engines can even be throttled (power output raised and lowered). A disadvantage of liquid propellants is that LH2 and LOX are potentially dangerous cryogenic (very low temperature) substances that pose severe storage and handling problems. Oxygen liquefies at -316F while hydrogen liquefies at -423F, just 37F above absolute zero (-460F). Liquid hydrogen is very volatile and hard to contain while liquid oxygen reacts violently when in contact with any hydrocarbon.

Solid fuel propellants have the fuel and oxidizer mixed together in granular form. The grains burn with an explosive power that is greater than gunpowder. The grains are packed inside a cylindrical shaped casing. There is a hollow core extending down the length of the grain pack in which the combustion occurs. The cross-section shape of the core can be configured to produce the desired thrust versus time curve. The big advantage of using solid propellants is that the engine is simple. It doesn't require large storage tanks with elaborate pumping and plumbing systems. The disadvantage of using solid propellants is that they don't burn as efficiently as liquid propellants and they can't be stopped and restarted in space. They essentially burn until all the fuel is consumed.

Another advantage of using liquid propellants over solid propellants is that liquid propellants have a higher "specific impulse" than solid propellants. Specific impulse can be defined as how long a rocket engine can produce a given amount of thrust using a given amount of fuel. Pound thrust is a measure of how powerful a rocket engine is while specific impulse is a measure of how efficiently it operates. An analogy can be made to an automobile engine where horsepower is a measure of engine power and miles per gallon is a measure of how efficiently it operates. In an automobile, a high powered engine is not too practical if it gets only a few miles per gallon of fuel. The specific impulse of some of the more advanced solid propellants is 265 seconds while the specific impulse of RP1/LOX is 330 seconds and the specific impulse of LH2/LOX is 450 seconds.

Missile Types

MISSILES ARE CLASSIFIED ACCORDING TO range, type of propellant, and number of stages. A stage is an independent engine/fuel tank section that after its use is jettisoned (separated) from the missile during flight. Most missiles have one to three stages. The first or lower stage is ignited at launch and burns until all its fuel is consumed. It is then jettisoned and the second stage ignited, etc. Generally, each succeeding stage is smaller than the preceding one. Staging is a means of keeping the missile thrust-to-weight ratio high enabling the accommodation of heavier payloads. (See Figure 3.) There are a number of ways that are used to classify missiles. Following is the one used in this publication that has been established by the U.S. Department of Defense.

• Intercontinental (ICBM) - over 5000 kilometers (over 3400 miles)
• Intermediate (IRBM) --- 3000-5000 kilometers (1860-3400 miles)
• Medium (MRBM) ------- 1000-3000 kilometers (0600-1860 miles)
• Short (SRBM) ----------- under 1000 kilometers (under 0600 miles)

The Atlas, Titan, and Minuteman ICBMs; the Thor and Jupiter MRBMs; and the Redstone SRBM are among the topics discussed in this publication.

Figure 3
Three Stage Missile
Saturn V Moon Rocket

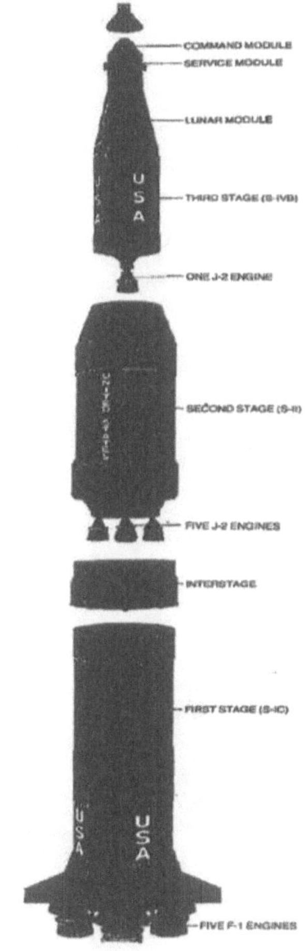

Missile Launch Sites

Cape Canaveral, Florida is where most of the military and civilian missiles are launched. Located on a strip of land that juts out into the Atlantic Ocean, Cape Canaveral is an ideal location to establish missile launch pads as missiles are launched east over the ocean making it safe for communities on the mainland. Also, missiles launched from the east coast of the United States near the equator get an extra boost of speed from the rotation of the Earth. The tangential velocity of the Earth at the equator equals approximately 1000 miles per hour. Missiles launched to the west over the Pacific Ocean lose velocity from this effect. Vandenberg Air Force Base in California has been used to launch missiles northward into polar orbit where the spinning of the Earth has little or no affect on velocity. (See Figure 4.)

The Atlas missile which was the first ICBM developed by the U.S. occupied Cape Canaveral launch complexes 11, 12, 13, and 14. Cape Canaveral launch complexes are located in a line north and south along the Atlantic Ocean coast. Each Atlas complex includes a test stand upon which the launcher is located, a "Block House" which is a domed structure with six feet thick concrete walls where engineers remotely control the various missile systems during launch countdown operations, and a "Ready Room" where the engineering offices are located.

The test stand is located two stories above ground level. There is a ramp leading up to it where missiles and equipment are transported to and from the launcher. There is a ten story high service tower that when fully deployed completely envelopes the missile giving engineers and technicians complete access to all mechanical and electrical systems most of which are located in pods on each side of the Atlas missile. The service tower is mobile and is rolled away from the missile before launch. There is a small tower with a pole at the top that supports the electrical "umbilical cord" through which information is sent to and received from the missile during countdown operations. The pole at the top is pivoted away from the missile just before launch, disengaging the umbilical cord plug from the missile. The "flame deflector" is a very large 90 degree duct just beneath the missile rocket engines. The duct deflects hot engine exhaust gases after engine ignition from the vertical to the horizontal just above ground level. The flame deflector walls are hollow and are filled with high pressure water from a 36 inch line. The inner walls of the flame deflector have many closely placed orifices through which water is jetted, cooling the engine exhaust gases and protecting the flame deflector from damage. (See Figures 5, 6, & 7.)

Figure 4
The Earth's Rotational Affect
(Looking Down From the North Pole)

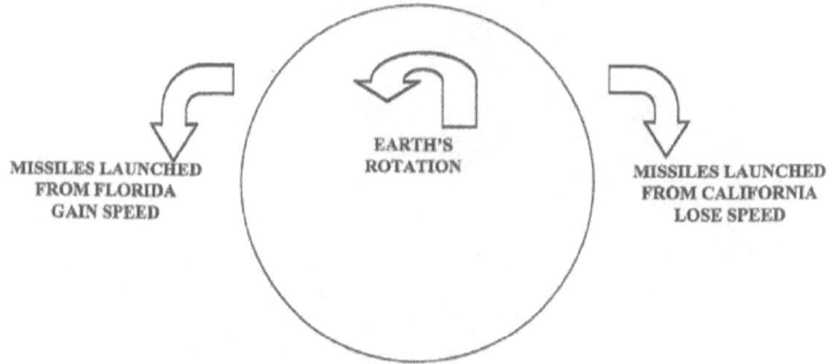

Figure 5

Figure 6
Schematic
Flame Deflector
Water Cooled

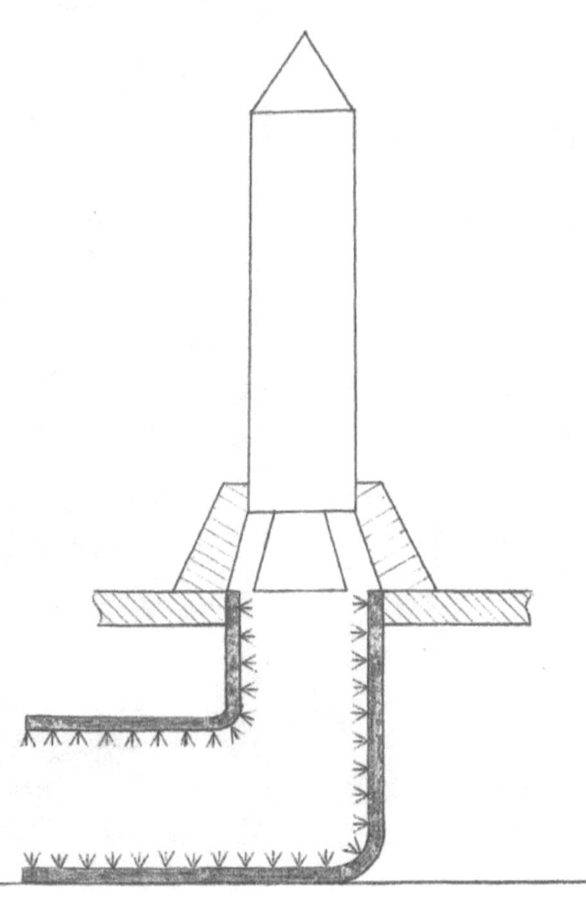

Figure 7
Atlas Missile Night Launch

The Atlas ICBM

As previously discussed, in October 1945 at the onset of the Cold War the U.S. Army Air Force requested industry proposals for a missile system to deliver a warhead as far as 6000 nautical miles. Six years later, after the outbreak of the Korean War, the Consolidated Vultee Company (Convair) of San Diego, California was awarded a contract to investigate the development of a ballistic missile system later named by Convair as "Project Atlas". In January 1955, the Air Force, under intense international pressure because of Russia having exploded both atomic and hydrogen bombs, ordered the Atlas into production even though it was only in the early stages of development.

The design that evolved was an ultra-thin skin airframe missile with integral propellant tanks that required 5 psi of internal pressure at all times to prevent it from collapsing under its own weight. (See Figure 8.) Wernher von Braun, the famous German rocket scientist, worried that the design would not survive the extremely high stresses at launch when all engines are operating under full thrust. The design included three RP1/LOX engines all of which were ignited at launch as in-flight engine ignition systems were not yet perfected. The two outside engines called "boosters" burned for a few minutes and then were jettisoned. The center engine called the "sustainer" continued to burn for a few more minutes and remained attached to the missile. All three engines were "gimballed" or pivoted at the base and were controlled by signals sent from the guidance system to steer the missile during powered flight. After that, two small engines called "verniers", one on each side of the missile, did the steering. The design incorporated a "nose cone" that separated from the missile and continued toward the target on a ballistic (arced) trajectory. The Atlas is sometimes referred to as being a one and one-half stage missile because of not having two separate sets of fuel tanks to go along with the two separate booster and sustainer engine stages.

The initial launch test versions were identified as "Atlas A" and were 75 feet, 10 inches long and ten feet in diameter. They were equipped with just the two booster engines and had a range of 600 nautical miles which was all that was needed to validate major design objectives such as the launch system and the thin-skinned airframe. There were a total of eight Atlas A launches made down the Air Force Eastern Range which extends from Cape Canaveral to the tip of Africa. See Figure 9. The following chart lists the results of the eight flights:

S/N	Date	Pad	Results
04A	06/11/57	14	Booster fuel system failure
06A	09/25/57	14	Booster fuel system failure
12A	12/17/57	14	600 mile success
10A	01/10/58	12	600 mile success
13A	02/07/58	14	Flight control failure
11A	02/20/58	12	Flight control failure
15A	04/05/58	14	(Records conflict)
16A	06/03/58	12	600 mile success

The launch system, missile thin-skin structural integrity, and other systems were considered validated by the tests; however, only three or four of the eight Atlas A missile launches were considered successful depending on which historical record is used. Regardless of which record is used, the performance of the Atlas A missiles left much to be desired since the system was released for production over two years before flight testing was started during a time in the Cold War when it was urgently needed.

The Atlas B test missiles were built more to the final design configuration with all three engines installed. The flights were made longer to test major events such as booster engine shutdown and separation, sustainer engine shutdown, vernier engine shutdown, and nose cone separation. The first Atlas B was launched from Cape Canaveral in July 1958. At approximately three quarters of a minute after launch, the engines lost power and the missile exploded. A total of ten Atlas B missiles were launched from Cape Canaveral, the last being six months later. Three of the ten launches ended in failure which was a record only slightly better that the Atlas A launches. Fortunately, an Atlas B missile was successful in December 1958 putting into earth orbit a famous message from President Eisenhower calling for world peace. The launch also put the U.S. ahead of Russia in the amount of weight put into Earth orbit. The Russian Sputnik launched in October 1957 was the first object put into Earth orbit by man. What worried U.S. officials was not only the Sputnik but the fact that the missile that put it into space was powerful enough to launch an H-bomb on the U.S. At the urging of the military, President Eisenhower accelerated America's missile programs. Within six months after Sputnik, the nation's space budget went from 0.5 to 10.5 billion dollars a year.

The Atlas C was more closely configured to the production version with each booster engine producing

165,000 pounds of thrust, the sustainer engine producing 57,000 pounds of thrust, and each vernier engine producing 1000 pounds of thrust. The missile was 6 feet, 8 inches longer than the Atlas B. Four of the six launches were successful with the last one held in August 1969 travelling the 6000 nautical miles distance down the Eastern Range.

The Atlas D was a prototype of the operational Atlas. As previously mentioned, the Air Force had released the Atlas for production even before vehicle testing was started because of the competition with Russia for leadership in missile development and deployment. The Atlas D was similar to the Atlas C except that each booster engine thrust was increased to 183,500 pounds. The first three test launches attempted between April and June 1959 all ended in failure; nevertheless, after two straight successes the missile was declared operational in September 1959. It was deployed in hardened bases in California, Nebraska, and Wyoming. Seven of the next ten launches were successful. It went on to become a reliable space exploration vehicle with names such as Atlas Able, Atlas Agena, Atlas Centaur, and Mercury Atlas which put the first U.S. astronauts into Earth orbit. The Atlas Able was used to launch a probe to the moon. The Atlas Agena was used in the Ranger program obtaining the first close-up images of the Moon and for the Mariner program which was the first spacecraft to fly by another planet. The LH2/LOX Centaur was put on top of the Atlas for dozens of launches such as the Surveyor Lunar Landing spacecraft and most of the Mars Mariner programs. (See Figure 10.)

The Atlas E was an upgraded version of the Atlas D. It had more powerful engines and an all-inertial guidance system. Previous Atlas missiles had guidance systems that relied on radio commands from the ground to aid in navigation. All-inertial guidance systems use signals from gyroscopes that remain in a fixed position relative to the position of the missile to steer the missile on its correct path. In July 1961 an Atlas E missile was successfully launched from Cape Canaveral for a distance of 9054 miles. It was deployed horizontally in hardened underground shelters in Washington, Kansas, and Wyoming.

The Atlas F was the final and most advanced of all the military Atlas ICBMs. It was modified to be stored vertically underground fully loaded with RP1 fuel. At launch, LOX was loaded, the missile raised, and then fired. This could be done in ten minutes, five minutes faster than the previous Atlas E. It was deployed in Kansas, New York, Nebraska, Oklahoma, Texas, and New Mexico.

The last of the one and one-half stage models was Atlas II which was used for non-warhead missions only. It had 63 successful flights with the last one held in August 2004. It is considered one of the most reliable launch vehicles in the world. With the Centaur upper stage and four strap-on boosters, it can put very heavy payloads into geosynchronous (fixed in the sky) and other Earth orbits.

The latest version of the Atlas is the Atlas V. It is an Atlas in name only. It no longer has thin skin and one and one-half staging. It incorporates a rigid frame for ease of transporting and handling. Ironically, the Atlas, originally designed to be used against the Soviet Union now uses Russian built engines for the first stage. It was put into service in 1999 and is used to put military payloads and commercial satellites into orbit.

Figure 8
Schematic
Atlas Missile
Integral Propellant Tanks

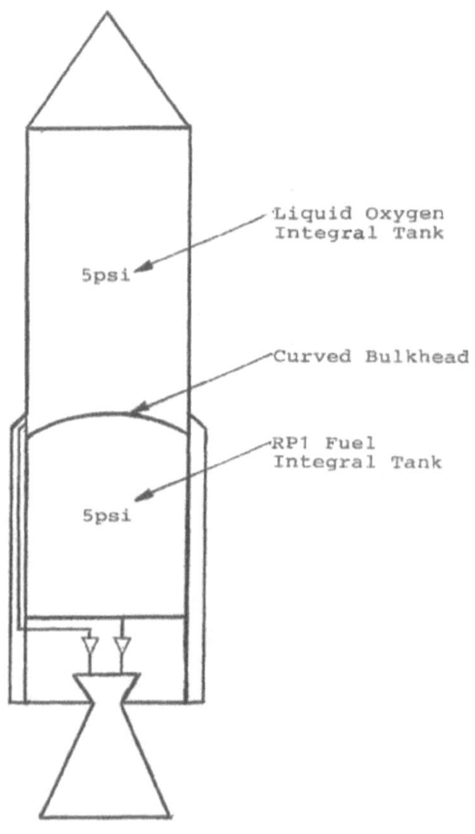

Figure 9

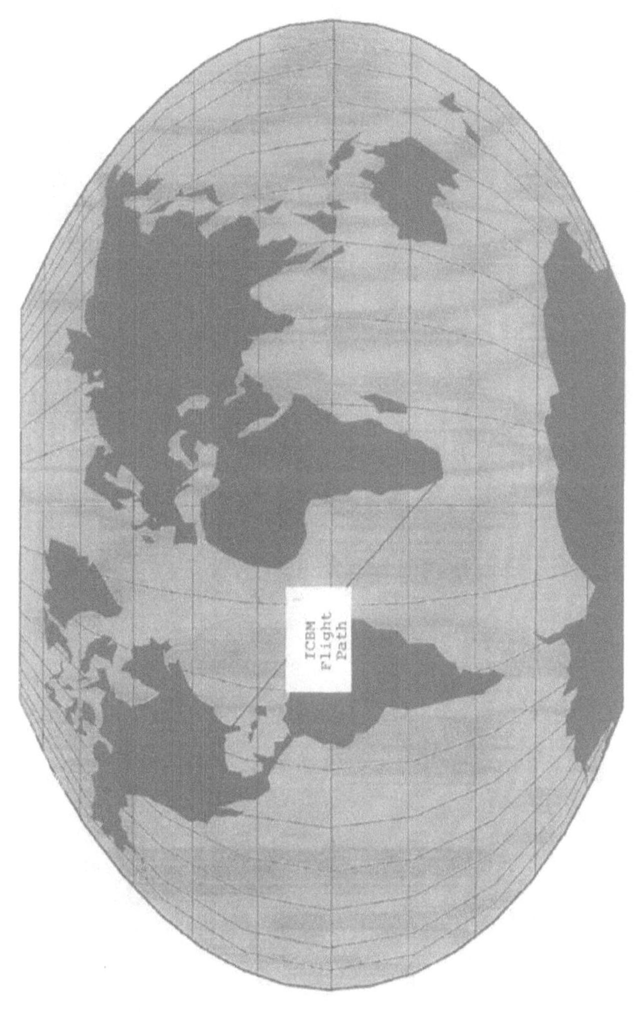

Figure 10

Atlas Mercury
Atlas Agena
Atlas Centaur
ESCAPE TOWER
MERCURY CAPSULE
LOX
FUEL
GUIDANCE
MAIN ENGINE
DROP-OFF BOOSTERS
GUIDANCE
SECOND STAGE
GUIDANCE
FUEL
LOX
SECOND STAGE
ATLAS MERCURY—THRUST: 360,000 LBS.
ATLAS AGENA B—THRUST: 360,000 LBS.
ATLAS CENTAUR—THRUST: 360,000 LBS.

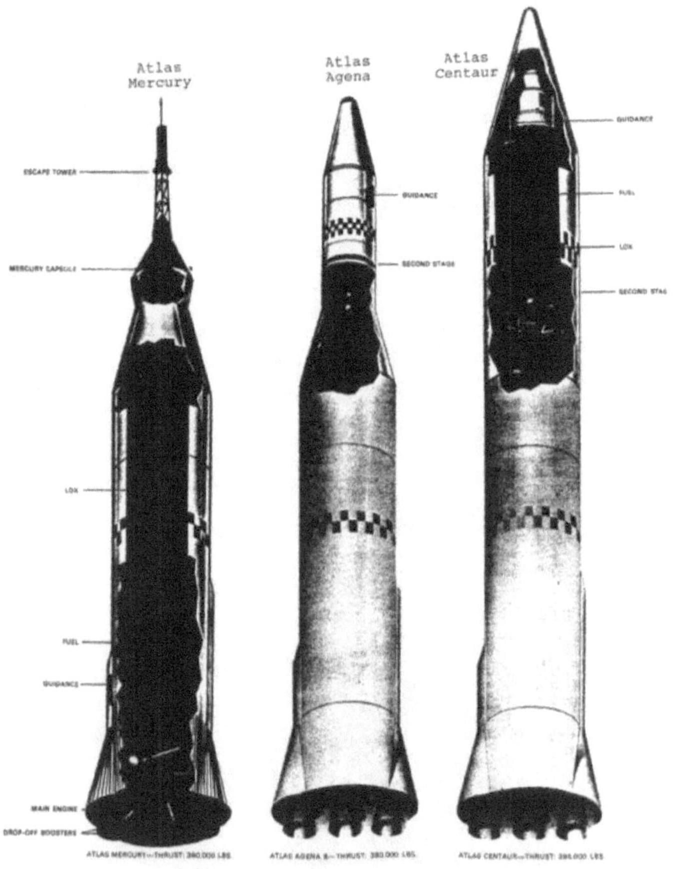

The Titan ICBM

THE TITAN WAS THE SECOND U.S. ICBM and the first to have two stages. The program began four years after Atlas. The Titan was initiated because of the possibility that the radical design approach of the Atlas would not be successful. The Titan I was a more conventional design with a rigid airframe compared to the internally pressurized thin-skin Atlas. The Titan I used the same propellants as the Atlas: RP1 and LOX. The first stage had 300,000 pounds of thrust while the second stage had 80,000 pounds of thrust. There were 70 launches of the Titan I dating from February 1959 to March 1965. Fifty-three were classified as successes and 17 as failures. Like the Atlas, the program was plagued with early failures but it finished with a high degree of success. The Air Force accepted delivery of the first Titan eight months before testing started. Two months after testing started, construction began on the first hardened site in Colorado. The Titan I was operational from 1962 to 1965 at which time Titan II and Minuteman missiles started their deployment.

The Titan II was designed with a first stage thrust of 430,000 pounds and a second stage thrust of 100,000 pounds; a big improvement over Titan I. Another improvement was the conversion to a hypergolic (self-igniting) fuel and a non-cryogenic oxidizer. This enabled the propellants to be stored in the vehicle at the launch sites and the missile fired within minutes. It also provided for a reliable second stage in-flight ignition. Titan II had an all-inertial guidance system thus not having to rely on ground signals for navigation. In February 1963, a Titan II launched from Cape Canaveral completed a successful 6500 nautical mile flight. With extended range, Titan IIs could now be deployed further south in the U.S. dispersing them more evenly across the states. The Titan II served to defend this country until 1981 when all sites were deactivated in favor of Minuteman. In the mid 1960s the Titan II was used in the NASA's Gemini two-man space program. All ten of the two-man capsules were successfully launched. In the late 1980s, the Titan II was used to launch a defense meteorological satellite from Vandenberg Air Force Base in California.

Titan III and Titan IV were modified Titan IIs with optional strap-on solid fuel boosters. They were used for launching military satellites into orbit as well as for NASA's Voyager probe to the outer planets, Viking probes to Mars, and the Cassini probe to Saturn.

The Minuteman ICBM

THE MINUTEMAN IS A THREE stage solid propellant ICBM that followed the Atlas and Titan into production and is the only ICBM deployed in the U.S. today. It was made possible because of advancements in solid propellant chemistry and engine designs that were developed since the conception of the Atlas and the Titan missiles. The advancements included more powerful burning propellants, engines with swivel nozzles for steering, and improved engine shutdown methods. The Minuteman is smaller, less complex, and much easier and faster launched from a hardened site than its predecessors and still able to carry a warhead a long distance.

The Minuteman missile is the brain child of Lieutenant Colonel Edward Hall of the Air Force Ballistic Missile Division. Hall, in 1957, generally working alone except for assistance from the consulting firm of Ramo-Wooldridge, incorporated new design features that were developed by a series of Air Force studies. It offered what hoped to be a missile that would put the U.S. ahead of Russia in the arms race. The Minuteman was first funded in early 1959, approximately three years after Titan. A test run in late 1959 proved that Minuteman could be launched directly from an underground silo without being raised to the surface saving precious minutes over Atlas and Titan. Also, the Minuteman propellants are an integral part of the missile while Atlas and Titan had to be loaded with fuel before launch. In February 1961 at Cape Canaveral, the first Minuteman was launched. All three stages performed flawlessly and the nose cone splashed down in the Atlantic Ocean 4600 nautical miles down range; a feat that Atlas and Titan did not come close to attaining. By the time the flight took place the Air Force was already planning deployment. The first bases were constructed in Montana, North and South Dakota, Missouri, and Wyoming.

Minuteman II with improved range, heavier payload, and greater accuracy replaced Minuteman I from September 1965 to March 1973. Minuteman III, an awesome weapon, carried a payload called MIRV (multiple independently targeted re-entry vehicle). One missile could deliver hydrogen bombs to multiple targets. By July 1975 there were 550 Minuteman IIIs and 450 Minuteman IIs deployed in the U.S., a force to be reckoned with for nearly 20 years. The fall of the Berlin Wall in November 1989 marked the beginning of the end of the Cold War. As a result of an agreement with Russia, the U.S. was to convert all Minuteman MIRVs to standard single warhead Minuteman IIIs. Five hundred single warhead Minuteman IIIs were to be deployed through 2020.

The MRBM and SRBMs

Because of the longer time needed to develop ICBMs, the U.S. developed the MRBM Thor and Jupiter missiles for fast deployment in Europe to counter the threat of Soviet Union ballistic missile programs. The Thor began development in January 1956 and within three years was deployed in the United Kingdom where it remained until 1963 after which time it was withdrawn in favor of the U.S. deployed ICBMs. The Jupiter program was started by the Army in late 1955 and became the first operational MRBM for the U.S. but was never deployed by the U.S. in favor of the Thor. The Jupiter was deployed in Europe by NATO until April 1957. Both were single stage MRBMs with one RP1/LOX engine developing 50,000 pounds of thrust. The Thor is used today as a space launch vehicle called the Delta. The Jupiter was used as a satellite and space probe vehicle called the Juno II.

The SRBM Redstone program was first started in 1953 and was a direct descendant of the German V2 rocket. It was active as a battlefield weapon from June 1958 to June 1964. It was designed as a surface-to-surface missile and was deployed by the U.S. Army in West Germany. The Jupiter A and Jupiter C missiles were actually modified Redstones. America's first satellite and first man in space were launched by a Jupiter C missile. (See figure 11.)

It can be said that the global expansion dreams of the Soviet Union, as described by Nikita Khruschev when he said, referring to the United States, "We will bury you!" were thwarted by the enormous effort put force by the U.S. military and industrial complex in developing spacecraft that ensure the safety of the citizens of, not only the United States, but of all of the peace loving nations of the world. Today the spacecraft and the knowledge learned is not only being used to secure the peace but is used in making great strides in such areas as space travel, space exploration, man moon landings, mars landings, astronomy, communications, weather forecasting, and global positioning systems.

Figure 11

Redstone Missile
Alan Shephard First American in Space
Gus Grissom Second American in Space

Author's Narrative

WHEN I GRADUATED FROM COLLEGE, my first job was an Engineer-in-Training for the Jack & Heintz Company in Cleveland, Ohio. Jack & Heintz made, among other things, starters for aircraft engines for the U.S. military. As I remember, it was an air-operated device that was one of the best of its kind. One day, the Cleveland Plain Dealer newspaper headline read "Hugh Missile Explodes off the Coast of Florida". It wasn't long after that I packed my belongings in my automobile and drove down to Cape Canaveral. I had already contacted Convair Astronautics (who later became General Dynamics Astronautics) and was offered a job on the Atlas missile program. I first rented a room in a private home on the mainland in Cocoa, Florida. A short while later I moved in with two other engineers, Bill Baker and Marvin Roberts, both southerners; Bill from Virginia and Marvin from Georgia. We lived in a rental home on Merritt Island. Merritt Island lies between the mainland and a strip of land on the ocean. The home was rather rustic to say the least. It was surrounded on three sides by a wooded swamp. When cutting the lawn, one had to wear high boots for fear of being bitten by a water moccasin. After a short stint there, I moved in with Bill Baker who had purchased a new home on Cocoa Beach which is on the Atlantic Ocean. The longer I lived in Florida, the more I headed toward the beach.

My first job with Convair was working in a hangar checking out various pneumatic pieces of missile ground support equipment before they were shipped to the launch complexes. Convair had four launch complexes: 11, 12, 13, and 14. Complexes 12 and 14 were completed first followed by 11 and 13. The one piece of ground support equipment that I remember well is the PCU which stands for Pressurization Control Unit. It was a maze of various components and associated plumbing encased in a sheet metal body and was about the size of an automobile van. It was installed on ground level next to the launcher and was to insure that the missile propellant tanks maintained 5 psi of pressure at all times. The Atlas missile was like a big balloon. It was constructed of very thin pieces of stainless steel and required pressure in the fuel tanks (which were integral with the skin) at all times to prevent it from collapsing on itself. Without internal pressure, it would "implode" as one Lead Engineer would say. Why it took so many different components in the PCU to accomplish this task, I don't know. There had to be more than one back-up system in that maze. One feature that it did have was to sound a siren if the tank pressure fell outside the prescribed limits. I don't remember if when the siren sounded we were to run for our lives or do something to restore tank pressure! The Atlas was basically two large fuel tanks with three rocket engines at the bottom and a nose cone on top.

While I was working in the hangar, something happened that you never read about in the newspaper. The first Atlas A missile exploded shortly after launch (the same one that I read about in the paper, I remember one engineer, Ken Stewart, saying "We were surprised that it even got off the pad"). Gold plated material was being sent to the hangar from our home office in San Diego and was being used to cover the bottom of the Atlas A test missiles. Initially, the bottoms of the missiles were partially open and now they were being completely covered with gold plated material which was supposed to act as a heat shield. It was thought that at higher speeds and altitudes rocket engine flames were being sucked back into the engine compartment and destroying critical components. This may not have been the only problem but was considered one of a number of things that may have caused the failure of the first missile that was launched. All subsequent Atlas A test missiles that I saw were wearing gold heat shields.

There was another thing that happened that you never read about in the newspaper that could have caused the failure. After my stint at Cape Canaveral, I worked at TRW, Cleveland, Ohio, on aircraft and missile design under the tutelage of Mr. Thomas Barish. Tom had one of the most brilliant engineering minds this country has ever seen. He told me that he was called to view Atlas propellant pumps being tested in a vacuum chamber. The test revealed that lubricating oil that was jetted to crucial rotating parts of the pump was never reaching its target because of "froathing" due to the reduced pressure at altitude. As I remember he told me the fix was to pressurize the pumps.

I was not happy in the hangar. I wasn't the type to work on a "behind the scenes" job. I wanted to be on the firing line. One day I got my wish and joined a crew that was overseeing the construction of one of the new Atlas launch complexes. What a breath of fresh air that was! One of my first assignments was to check out the missile "Hold Down and Release System". There were two large pieces of equipment called "heads", one on each side of the launcher at the base of the missile. Each one had a large pin that engaged a socket in the missile. At launch, after the engines were successfully ignited and delivering full thrust, pressure was released in the heads and a mechanism driven by the upward motion of the missile retracted the pins freeing the missile for flight. There was a test that was to be run on the system called the "Blowdown Test" before it could be declared operational. The release of pressure from the heads was controlled by an orifice and the pressure versus time plot had to be within certain limits. For the longest time I couldn't get the correct plot. One day I decided to inspect the valve on the launcher which controlled the release of pressure and noticed that it had markings on it indicating how the lines were to be connected. I noticed that two of the lines were criss-crossed. I brought this to the attention of my boss Mr. Tom O'Malley. (Tom became a Test Conductor and was the person who pushed the button that sent John Glenn on his way to Earth orbit.) There was a big argument between Tom and Lead Engineer Bob Lindsey over the validity of my claim. We won the argument. The lines were changed and the system operated as designed from then on.

After the complex was completed, our group was made a launch team. My first assignment was "Test Stand Engineer". Among other things, I had the PCU (which I already explained controlled missile fuel tank pressure), the RVU or Relief Valve Unit, the LBU or Launcher Booster Unit, the Purge Box, the Stabilization System, the Hold Down and Release System (already explained), and the various firefighting systems. These systems, for the most part, were a maze of dynamic components and lines. One person could not be expected to know what the function of each part of every system was. We knew how to check the entire system out but if there was a major problem Design Engineering from California was called and it wasn't long before personnel were sent to the Cape to fix it.

Tower Firex was a fire fighting system that incorporated a ring of nozzles completely surrounding the missile at each level of the ten story high service tower. All the fire fighting system water lines ran through an excavation in the ground called the "Valve Pit". There was a motorized shutoff valve and a manual shutoff valve in series in each water line. The motorized valve was made to open and close from a control panel in the Blockhouse. Normally the manual valve was left open and the motorized valve closed and if a fire occurred, the motorized valve was opened from the Blockhouse panel switch releasing pressurized water to fight the flames. One of the pre-launch checks was to insure the proper functioning of the motorized valve by closing the manual valve and cycling the motorized valve open and closed from the Blockhouse. One day, while in the process of running this test, I called over my headset to "Sarge", a technician who was stationed in the valve pit to close the manual valve in the Tower Firex System so I could check the operation of the motorized valve. After a few minutes I received the message over my headset from Sarge "Tower Firex manual valve closed". I then switched the Tower Firex motorized valve switch to the "Open" position. Seconds later I received a very loud shout over my headset; "turn off the @*!&# water"! Sarge had closed the wrong manual valve and the missile and people on all levels of the service tower were being drenched! I immediately switch the Tower Firex switch to "close". It seemed like it took forever for the valve to respond! A few minutes later my boss, Jim Vivera, came running into the Blockhouse and told me to report to the Ready Room immediately to explain exactly what happened. I went into the office and accepted full responsibility for the incident even though it was Sarge who let me down. As I sat there on the hot seat, it was almost humorous as the engineers who were working on the missile at the time came into the room. Some were drenched and had water running down their noses while they were cursing me out. I remember asking one engineer who was working on the missile at the time what his reaction was when the water hit. He said he immediately slammed the pod door shut and ran. The Atlas missile has pods on each side where electronic and mechanical equipment is contained. To this day I think how quick acting this engineer was to slam the pod door closed before running away. I believe there were a few more who did the same because the missile was dried out and successfully launched the next day. Needless to say, we never ran Tower Firex tests again with a missile in place.

The Launcher Booster Unit was a piece of equipment that was about as big as a large upright piano. It was located in a room just below the missile and provided pressure for the various launcher systems. During a night countdown prior to a launch just after the missile was completely loaded with fuel and liquid oxygen, there was a momentary power failure. Everything switched back on except the Launcher Booster Unit which had to be operating for the missile to be launched. I volunteered to leave the protection of the Block House, go down to the test stand, crawl up a flight of stairs, and in the dark manually turn on the LBU which was located directly under the hissing and steaming missile that was minutes away from being launched. You can bet that it wasn't long before that problem got fixed and never happened again!

The Purge Box was a rather small unit, comparatively speaking, about the size of a suitcase that sat on the launch pad near the base of the missile. It contained a number of valves that were used to supply fluid to clean the various lines and parts of the rocket engines after static firings and before launches. There were certain non-metallic seals in the valves that were found, after the fact, not to be compatible with all the engine cleaning fluids. Occasionally a valve leaked and seals had to be changed. During the middle of a rainy night on the day before a launch Tom O'Malley drove his VW Beatle right up to our front door and sounded the horn. I climbed out of bed, went to the door, and heard Tom say "Purge Box". I knew exactly what he wanted, dressed, and drove to the cape to change Purge Box seals. Tom, as I mentioned is the person who pressed the button to send John Glenn on his way to become the first U.S. astronaut to orbit the Earth.

I had another night experience that deserves mention. I was on my way to work on second shift and when I arrived on the Cape I looked up and saw an Atlas missile a couple hundred feet off the ground coming right at me! I remembered that the complex next to ours had a launch scheduled for that time. What a launch it was! When the missile first moved vertically off the test stand one of the engine fuel and drain valves did not close. Fuel came pouring out one side at the base of the missile starving one engine and preventing it from providing thrust. The opposite engine operated normally steering the missile to a horizontal trajectory just above the ground. That's when I first saw it and it was just moments later that the Range Safety Officer sent the signal for the missile to self-destruct. It erupted into a huge ball of flames and set fire to some dry vegetation on the cape. The fire sent water moccasins scurrying out of the brush onto the roads. The security police shot them for protection to unsuspecting passers-by.

The Relief Valve Unit was an assembly about as large as two suitcases. It was located on the launcher near the base of the missile. It was used to vent the various gases that were used in and around the missile. I remember that we could associate the number of leaks in a line to the size of the molecules of the gas contained inside such as nitrogen and helium.

The Stabilization System was an elaborate network of hydraulic lines and valves that controlled the up and down motion of four pins, spaced 90 degrees apart, around the bottom perimeter of the missile. The purpose of this system was to keep the missile pointed straight up when it was sitting on the test stand no matter how hard and from what direction the wind was blowing. I remember looking at the schematic of the system and thinking what a masterpiece of engineering design it was. It was a completely passive system that relied on no outside energy source but nevertheless functioned flawlessly.

After a few launches, I was made Propellant Loading Engineer. We loaded fuel in the missile the day before launch and then we loaded the oxidizer during the countdown on the day of the launch. It was all done remotely controlled from the Blockhouse. Fuel loading the day before launch with RP1 is a relatively easy task compared to loading liquid oxygen. Liquid oxygen is stored at a very low temperature and is very dangerous as it explodes when it comes into contact with any hydrocarbon. The day of the launch at T minus 30 minutes, the LOX loading operation begins. I would start the operation at the Blockhouse panel by remotely starting two massive pumps that were used to rapid fill the missile tank from a ground storage tank. At about the 90% fill level which I could determine from a gage on the Blockhouse panel, I switched to a small variable speed pump to bring the liquid level to a few percentage points under 100% and maintained it there. A few minutes before launch, Test Conductor Wick Jackson would address all the Test Engineers regarding the status of their systems at which time they would reply "Go" or "No go". On one launch I could not fill the missile LOX tank above the 90% level with the small pump. When Wick Jackson gave his status check, I answered "No go"! To this day I can't imagine the amount of grief that I gave Wick Jackson for that "No Go". It takes so many hard days and nights to prepare a missile for launch and now it had to be postponed because a young engineer had the audacity to give a "No go" two minutes before launch for not being able to come within a few percentage points of filling the missile with LOX. During the next few days an engineer was sent from California who determined that the problem was with the gage. I still wonder if there was enough LOX in that tank to support a successful launch. Anyway, the problem was fixed and there were no further issues with LOX gauges. I was recently told by a friend who works at NASA's Plum Brook Test Facility in Sandusky, Ohio that he has trouble with the same type of variable capacitance LOX gauge. Also, I read liquid oxygen is now filled in a missile until it starts to flow out the Fuel Fill and Drain Valve which is located at the top of the LOX tank. This type of operation which apparently is now being used sounds crude but the level of liquid oxygen was, and probably still is, very difficult to attempt to measure in the missile tank.

Several more incidents happened while I was at the Cape that deserve mention. During one of our launches, everything was proceeding normally when late in the countdown a hold was called. A stranger went out to the missile and did something to it that most of us knew nothing about. The countdown was completed and the missile successfully launched. Afterwards, at our post launch gathering at the "Starlight Inn", it was announced that the missile successfully traveled over the continent of Africa and the nose cone landed in the Indian Ocean some 10,000 miles away, about twice the reported range of the Atlas. As it turned out, it was done as a show of force proving that the United States had the capability of reaching any target on the planet.

The launch team next to our complex put an Atlas missile into Earth orbit with the famous Eisenhower "Peace Message" broadcast from it during one Christmas season. By doing so, the U.S. took the lead in putting a heavier than Sputnik object into Earth orbit. On a clear and sunny day we could see the Atlas with its gleaming stainless steel skin pass overhead.

Members of my family were vacationing at the Cape one year when our crew launched an Atlas missile into the night sky. The missile started out beautifully but when it reached a point high in the sky it exploded. (That is when one our lead engineers shouted, "It imploded, it imploded!" referring to the fact that tank pressure was probably lost and with the propellant pumps still operating the missile collapsed inward.) My family members saw the good and the bad, a perfect launch for a few minutes then an explosion high in the sky. The seven original astronauts were in the Blockhouse at the time but the failed mission didn't seem to bother them even though one of them was to ride an Atlas into Earth orbit. I wish I would have known at the time that John Glenn, one of the astronauts and a fellow Ohioan was there; I would have shaken his hand!

One day we were told that we were going to launch a space probe to the planet Venus. It was called the Atlas Able. The window of time within which the missile was to be launched was very narrow. We were told to have all our systems in perfect order and that we were to have 100% of our spare parts on hand at all times. One crucial aspect was that the weight of the missile at launch had to be held within very close limits. I was sent to California to conduct special propellant loading tests as the propellants comprised a very large portion of the missile total weight. The plan devised in California was that I was to load propellants to a given weight rather than to a percent fill as was done in the past. There was to be instrumentation installed so that an accurate missile weight was known at all times. After the training in California I arrived back in Florida only to find that the mission was cancelled. I was so disappointed that I quit Cape Canaveral and returned back to Ohio. I later read in a magazine that the mission was changed to send Atlas Able to the moon for a soft landing but, when they tried, the landing was not so soft!

www.ingramcontent.com/pod-product-compliance
Lightning Source LLC
La Vergne TN
LVHW090025070726
842759LV00025B/270